The California Condor, "The Big Ugly"

MARY JO NICKUM

Aquitaine Ltd
Phoenix, Arizona

The California Condor, "The Big Ugly"

MARY JO NICKUM

Aquitaine Ltd
Phoenix, Arizona

Aquitaine, LTD

ISBN: 9781736467206
Library of Congress Cataloging Number LCCN: 2020950683

Printed in the United States of America

First Edition

DEDICATION

This book is dedicated to young people everywhere who
Don't like to or find it difficult to read and
Who, therefore, live on the fringes of a happy, healthy life.
Learn to enjoy reading and the world can be yours.

Believe in yourselves

Table of Contents

Fig 1. California Condor

Chapter 1

What is a California Condor?

The California condor, *Gymnogyps californianus* **(Figure 1)**, is one of the largest flying birds in the world. When it soars, the wings spread more than nine feet from tip to tip. Condors may weigh more than 20 pounds. The male Andean condor of South America is even larger than our California condor. Both are endangered species.

The adult California condor is a uniform black with the exception of large triangular patches or bands of white on the underside of the wings **(Figure 2)**. It has gray legs and feet, an ivory-colored bill, a frill of black feathers surrounding the base of the neck, and brownish red eyes. The juvenile is mostly a mottled dark brown with blackish coloration on the head. It has mottled gray instead of white on the underside of its flight feathers.

Fig 2. White beneath wing

There are few feathers on the head and neck of the condor **(Figure 3)** and the skin of the head and neck is capable of flushing noticeably in response to emotional state, a capability that can serve as communication between individuals. The skin color varies from yellowish to a glowing reddish-orange. The birds do not have true **syringeal vocalizations**. They can make a few hissing or grunting sounds only heard when very close.

Fig 3. Head and neck

Contrary to the usual among true birds of prey, the female is slightly smaller than the male. Overall length can range from 43 to 55 in. and wingspan from 8.2 to 9.8 ft. Their weight can range from 15 to 31 lb., **(Side Bar A)** with estimations of average weight ranging from 18 to 20 lb. Wingspans of up to 11 ft. have been reported but no wingspan over 10 ft. has been verified. Most measurements are from birds raised in captivity, so determining if there are any major differences in measurements between wild and captive condors is difficult **(Side Bar B)**.

California condors have the largest wingspan of any North American bird **(Figure 4)**. They are surpassed in both body length and weight only by the trumpeter swan and the introduced mute swan. The American white pelican and whooping crane also have longer bodies than the condor. Condors are so large they can be mistaken for a small, distant airplane, which possibly occurs more often than they are mistaken for other species of bird.

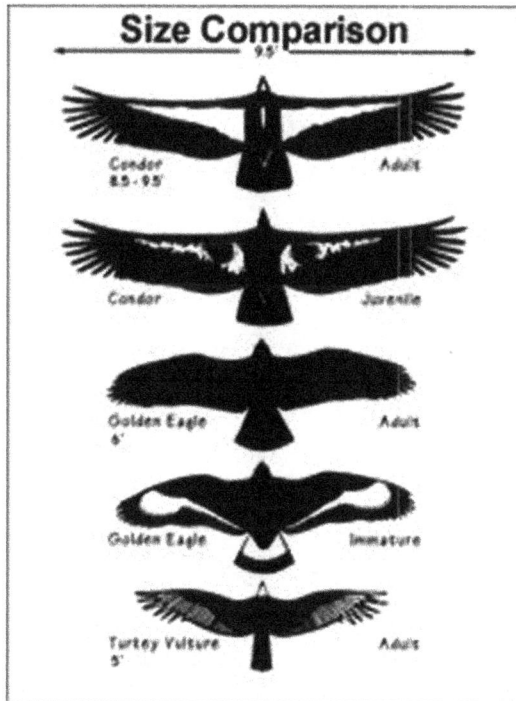

Fig 4. Size comparison

The middle toe of the foot of the California condor is greatly elongated and the hind one is only slightly developed **(Figure 5)**. The talons of all the toes are straight and blunt, and are thus more adapted to walking than gripping. This is more similar to their supposed relatives, the storks, than to birds of prey and Old World vultures, which use their feet as weapons or organs of **prehension**.

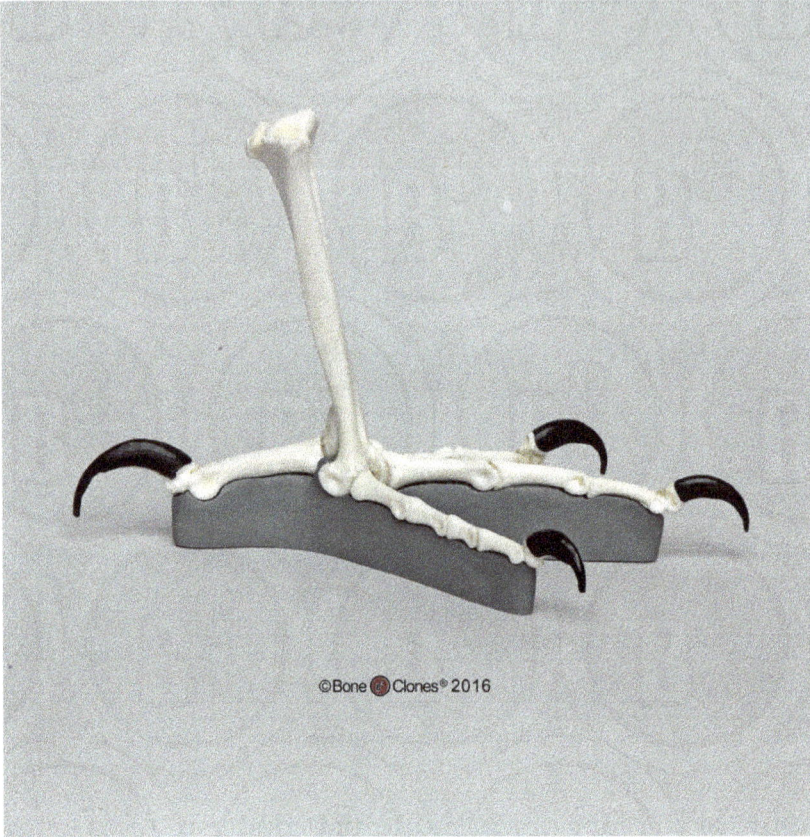

Fig 5. Condor foot

Chapter 2

History of the California Condor

During the Pleistocene **(Side Bar C)**, which began more than 250 million years ago and lasted until about 11,000 years ago, California Condors soared over parts of eastern, western, and southern North America along coastlines and in forests, mountain canyons, grasslands, and deserts. On the coast, they fed on dead seals, whales, and other marine animals that washed to shore. Inland, they likely feasted on the remains of large mammals, such as mastodons and mammoths. As these large animals, known as megafauna, began to disappear, so did California Condors.

Let us take some time to understand the Pleistocene epoch. Pleistocene is the period in Earth's history we commonly refer to as the Ice Age. Through much of this period, the Earth's northern and southern regions were covered by feet-thick glaciers **(Figure 6)**. It is important to recognize that the Pleistocene was a series of real ice ages, separated by relatively short interglacial periods. The Pleistocene started 2.6 million years ago and lasted until the termination of the **Weichsel glaciation (Figure 7)** about 11,711 years ago.

Side Bar C: Geologic Timetable

<u>**Precambrian**</u> **(4,500 to 543 mya*)**
   <u>Hadean</u> (4500 to 3800 mya)
   <u>Archaean</u> (3800 to 2500 mya)
   <u>Proterozoic</u> (2500 to 543 mya)
      <u>Vendian</u> (650 to 543 mya)

<u>**Phanerozoic**</u> **(543 mya to today)**
   <u>Paleozoic Era</u> (543 to 248 mya)
      <u>Cambrian</u> (543 to 490 mya)
         <u>Tommotian</u> (530 to 527 mya)
      <u>Ordovician</u> (490 to 443 mya)
      <u>Silurian</u> (443 to 417 mya)
      <u>Devonian</u> (417 to 354 mya) —The Age of Fishes
      <u>Carboniferous</u> (354 to 290 mya)
         Mississippian (354 to 323 mya)
         Pennsylvanian (323 to 290 mya)
      <u>Permian</u> (290 to 248 mya)

   <u>**Mesozoic Era**</u> **(248 to 65 mya)**
      <u>Triassic</u> (248 to 206 mya)
      <u>Jurassic</u> (206 to 144 mya)   The Age of Dinosaurs
      <u>Cretaceous</u> (144 to 65 mya)

   <u>**Cenozoic Era**</u> **(65 mya to today)**
      Tertiary (65 to 1.8 mya)
         <u>**Paleocene**</u> (65 to 54.8 mya)
         - Began with extinction of dinosaurs.
         Emergence of early mammals
         <u>**Eocene**</u> (54.8 to 33.7 mya) – Emergence of first modern
         mammals. Epoch ended with a major extinction event.
         <u>**Oligocene**</u> (33.7 to 23.8 mya) – A relatively quiet time for
         mammalian evolution, few new faunas appeared.
         <u>**Miocene**</u> (23.8 to 5.3 mya)
         - Recognizably modern mammals appeared

Pliocene (5.3 to 1.8 mya)
– Modern mammals continue to diversify
Quaternary (1.8 mya to today)
Pleistocene (1.8 mya to 10,000 years ago) – Includes the ice ages
Holocene (10,000 years ago to today) – Recent era

*mya—million years ago.

Fig 6. Ice Age glacier

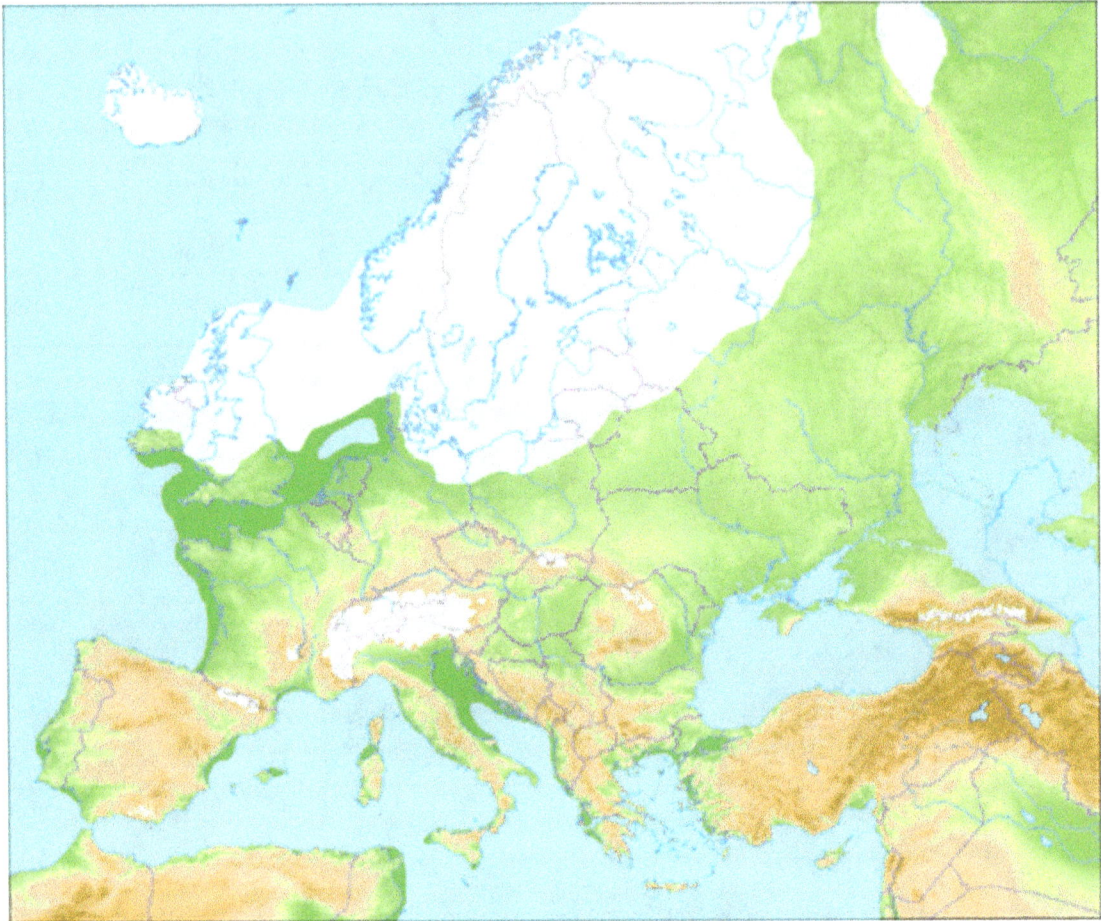

Fig 7. Europe during the last glacial period

In Earth's climate history huge glaciers had covered large parts of the northern and southern continents several times **(Figure 8)**. Already at the beginning of Proterozoic occurred the prolonged Huronian Ice Age, at the end of Proterozoic occurred three ice ages, namely Stuartian, Marinoan (also known as the Varanger Ice Age after the Norwegian peninsula, where it was first detected) and Gaskiers. In Phanerozoic came first the Andean-Saharan Ice Age (also called the Hirnantian Ice Age) at the transition from Ordovician to Silurian and later the Karoo Ice Age at the transition from Carboniferous to Permian. Compared to previous glacial periods, the Pleistocene Ice Age has, until now, lasted only a short time.

Fig 8. Ice sheet

The cold climate of the Pleistocene was a natural continuation of the last 55 million years of falling temperatures **(Figure 9)**. In particular two factors were essential for the formation of the great glaciers. One was the temperature dropped so much that the snow did not melt during the summer and, thus, could accumulate year after year. Second was Earth's continents were so positioned that warm ocean currents flowed against the north, and released their heat and moisture as precipitation in the form of snow.

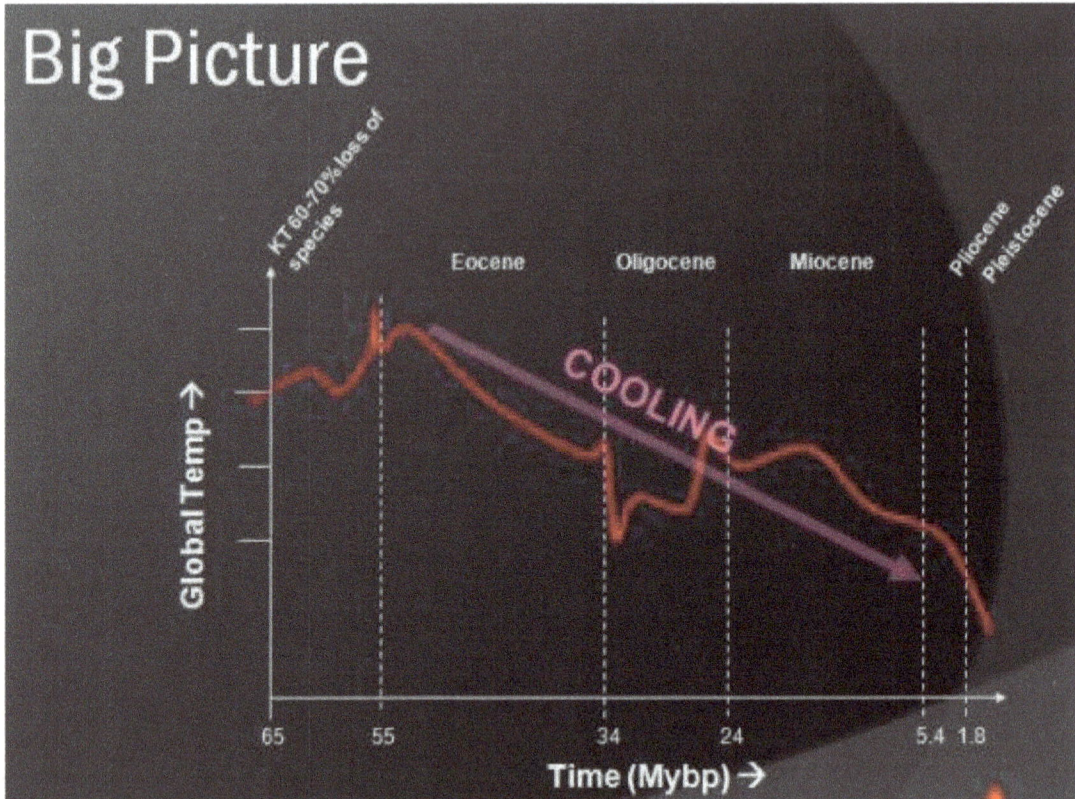

Fig 9. 55 million years of falling temperatures

The entire Quaternary period **(Side Bar D)** is often referred to as the ice age, because two large permanent glaciers continuously existed during the period, namely on Antarctica and Greenland. During the Pleistocene's coldest periods, which also are called ice ages, existed also enormous glaciers in Europe, North America and in Patagonia on the southern hemisphere. The shorter and warmer intervals between the recurrent Pleistocene glaciations are termed interglacials.

Over long periods of time, 30 percent of Earth's land masses were covered by dazzling white ice and snow and, thereby, Earth's **albedo** increased dramatically, and a very high proportion of solar radiation was reflected back to space, which intensified further cooling.

Quaternary Period

Eonothem/ Eon	Erathem/ Era	System/ Period	Series/ Epoch	Stage/ Age	millions of years ago
↑	↑	↑	Holocene	Upper	0.0117
					0.126
				Middle	
					0.781
Phanerozoic	Cenozoic	Quaternary	Pleistocene	Calabrian	
					1.806
				Gelasian	
↓	↓				2.588

The Geologic Epoc in which the California Condor lived, also known as 'The Ice Age'.

Geological congresses have agreed the Pleistocene ice age ended about 11,711 years ago **(Side Bar E)** and we have defined that present is a brand new period, called the Holocene; however, we cannot define and decide us away from the fact the present Holocene warm period is an interglacial, of which there have already been many. With high probability, the glaciers will return to the northern and southern parts of the world's continents, only we do not know when.

Side Bar E: The Pleistocene Epoch—The Rise of the Condor

Pleistocene Epoch, earlier and major of the two epochs that constitute the Quarternary Period of the Earth's history, and the time period during which a succession of glacial and interglacial climatic cycles occurred. The base of the Gelasian Stage (2,588,000 to 1,800,000 years ago) marks the beginning of Pleistocene, which is also the base of the Quarternary Period. It is coincident with the bottom of a marly layer resting atop a sapropel called MPRS 250 on the southern slopes of Monte San Nicola in Sicily, Italy, and is associated with the Gauss-Matsuyama geomagnetic reversal. The Pleistocene ended 11,700 years ago. It is preceded by the Pliocene Epoch of the Neogene Period and is followed by the Holocene Epoch.

The Pleistocene Epoch is best known as a time during which extensive ice sheets and other glaciers formed repeatedly on the landmasses and has been informally referred to as the "Great Ice Age." The timing of the onset of this cold interval, and thus the formal beginning of the Pleistocene Epoch, was a matter of substantial debate among geologists during the late 20th and early 21st centuries. By 1985, a number of geological societies agreed to set the beginning of the Pleistocene Epoch about 1,800,000 years ago, a figure coincident with the onset of glaciation in Europe and North America. Modern research, however, has shown that large glaciers had formed in other parts of the world earlier than 1,800,000 years ago. This fact precipitated a debate among geologists over the formal start of the Pleistocene, as well as the status of the Quaternary Period, that was not resolved until 2009.

Specialists defined the Weichsel glaciation as having ended about 11,711 years ago, and then the temperature on the surface of the ice was about 4 degrees lower than today. If we draw a horizontal line across Pleistocene's temperature graph **(Figure 10)** at 4 degrees lower than today's temperature, we will very roughly be able to separate the actual glacials from non-glacial periods by the same definition.

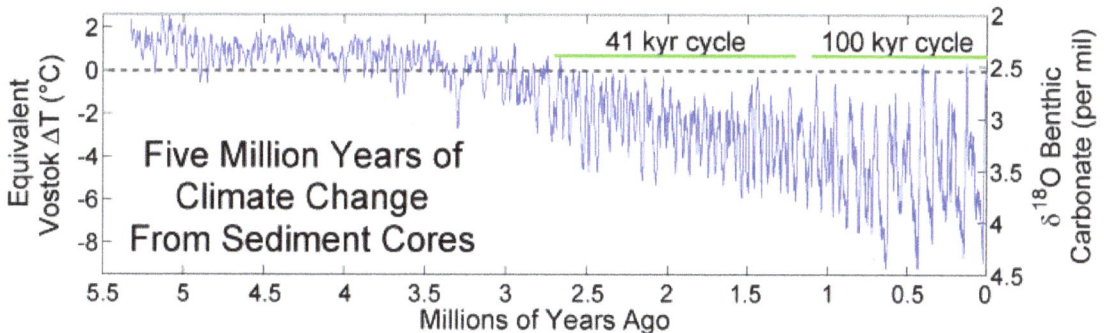

Fig 10. Climate change in the Pleistocene

In the early Pleistocene, it was cold, but not extremely cold. Relatively small glaciers periodically emerged, which may have been limited to northern Scandinavia, the Norwegian Mountains, northern Canada, Greenland and, most likely, some Arctic islands. The cycle time between cold periods and warm periods was 41,000 years. The temperature difference between cold and warm periods was about 4 degrees or less.

One can also see during the last part of the early Pleistocene, between 1.8 and 0.8 million years before present, it became somewhat colder and temperature variations between cold and warm periods became greater. More than half the time, the average temperature was lower than 4 degrees below the present temperature and, according to our rough definition, real glaciation periods. In glacial periods, the edge of the glacier in Europe stood, perhaps, along the Norwegian coast and the Swedish lakes. The cycle time between cold and warm periods was still about 41,000 years.

The landscape of northern Europe may have been **tundra,** today known in northern Russia; but, as mentioned above, we have no certain knowledge because 3,300 ft.-thick glaciers have since removed all traces **(Figure 11).**

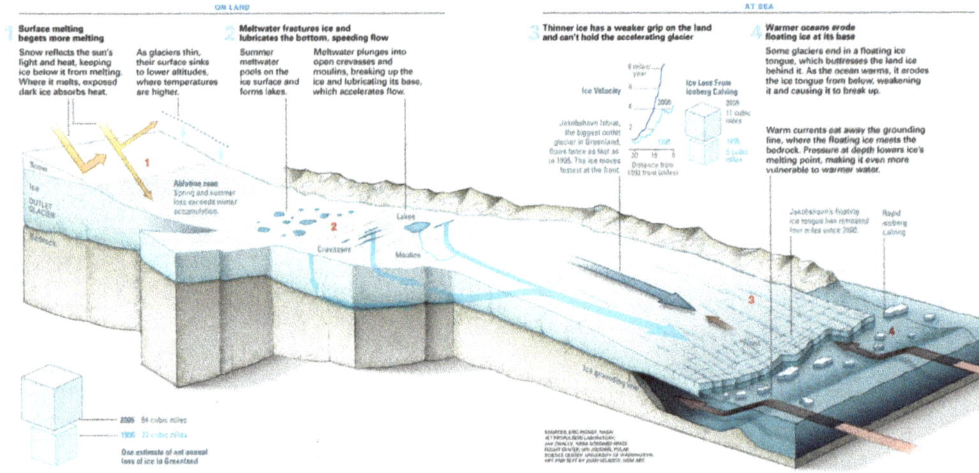

Fig 11. Ice melt

The Pleistocene was essentially a cold period, which was reflected in the fact all climate zones were shifted toward the equator compared to today. During Last Glacial Maximum tundra and mammoth steppe extended down to the Alps and the Mediterranean Sea was surrounded by a sparse stand of pine trees.

Climate zones on the mountain slopes were lower than today. In some places, the snowline could be almost 2,953 ft. further down than in the present. In Africa, the now ailing glaciers on Mount Kilimanjaro and on the Ruwenzori range between Uganda and Congo were greater. There were also glaciers on the mountains of Ethiopia and on the western Atlas Mountains.

The genus *Gymnogyps* (California condor) is an example of a **relict** distribution. During the Pleistocene epoch, this genus was widespread across the Americas. From fossils, the Floridan *Gymnogyps kofordi* from the Early Pleistocene and the Peruvian *Gymnogyps howardae* from the Late Pleistocene have been described in many other places. A condor found in Late Pleistocene deposits on Cuba was initially described as *Antillovultur varonai*, but has been recognized since as another member of *Gymnogyps*, *Gymnogyps varonai*. It may even have derived from a founder population of California condors.

Now, a dozen new radiocarbon dates, together with a thorough review of its fossil distribution, shed new light on the time and probable cause of extinction of the California condor, *Gymnogyps californianus*, in Grand Canyon, Arizona. The radiocarbon data indicate this species became extinct in the Grand Canyon, and other parts of the inland West, more than 10,000 years ago in coincidence with the extinction of megafauna (proboscidians, edentates, perissodactyls). That condors relied on the megafauna for food is suggested by the recovery of food bones from a late Pleistocene nest cave in Grand Canyon. These fossil data have relevance to proposed release and recovery programs of the present endangered population of California condors.

Today's California condor is the sole surviving member of *Gymnogyps* and has no accepted subspecies; however, there is a Late Pleistocene form that is sometimes regarded as a palaeosubspecies, *Gymnogyps californianus amplus*. Current opinions are mixed regarding the classification of the form as a **Chronospecies (Side Bar F)** or a separate species *Gymnogyps amplus*. *Gymnogyps amplus* occurred over much of the bird's historical range, even extending into Florida, but was larger, having about the same weight as the Andean condor. This bird also had a wider bill. As the climate changed during the last ice age, the entire population became smaller until it had evolved into the *Gymnogyps californianus* of today .

Side Bar F: The Condor Is a Chronospecies

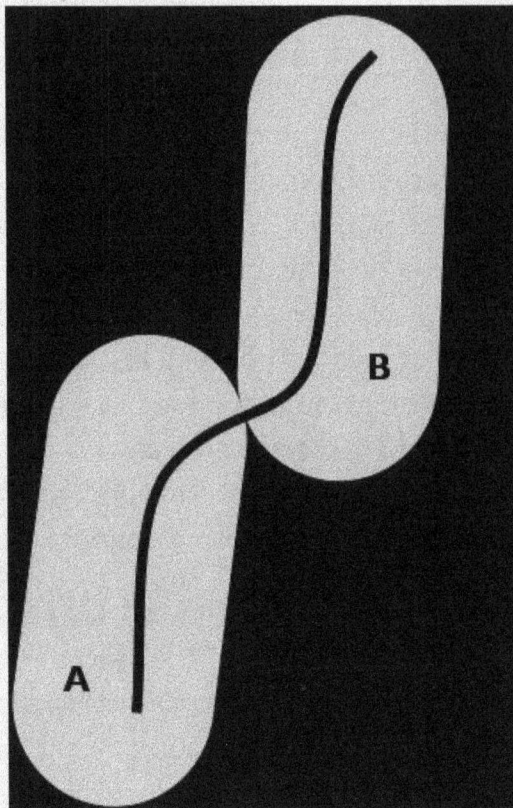

In **paleontology**, the evidence for species and evolution comes mainly from the comparative anatomy of fossils. A **chronospecies** is defined in a single lineage (solid line) whose morphology changes with time. At some point, paleontologists judge that enough change has occurred that two forms (A and B), separated in time and anatomy, once existed. If only sporadic examples of each survive in the fossil record, then the forms will appear sharply distinct.

Chapter 3

California Condor Diet

Condors are carrion eaters. Carrion isn't the bag you take with you on a plane – it means any animal that is already dead. Because condors don't hunt their food, they must travel far and wide in search of dead animals. They may make a meal out of an animal as small as a ground squirrel or as big as a cow. Some common animals on their menu include sheep **(Figure 12)**, deer, elk, and horses. In their lifetimes, condors may range over millions of acres in search of food.

Fig 12. Sheet as carrion

The swallow bone chips and marine shells meet their calcium needs. They favor small to medium-sized carcasses, probably because smaller bones are easily consumed and digested. Condors locate carcasses with their keen eyesight, not by smell, and by observing other **scavengers** assembled at a carcass. When they land, they take over the carcass from smaller species but they are tolerant of each other and usually feed in groups **(Figure 13)**. Condors are wary of humans while feeding, which is probably why they do not use roadkill as a food source.

Fig 13. Condor feeding on carrion

Paleontological evidence suggests populations of these obligate scavengers were associated with the carcasses of large animals. After the late Pleistocene extinction of most large terrestrial mammals in North America, condors appear to have been restricted to the west coast, where stranded marine mammals offered the only remaining abundant source of large animal carcasses **(Figure 14)**.

Fig 14. Marine mammal carcass

There is little direct evidence marine mammals were a major component of condor diets beyond scattered historical observations, however. In 1806, Lewis and Clark observed condors feeding on whales near the mouth of the Columbia River. Captain Clark (Figure 15) wrote on February 16, 1806: "This bird fly's very clumsily, nor do I know whether it ever seizes its prey alive, but am induced to believe it does not. We have seen it feeding on the remains of the whale and other fish which have been thrown up by the waves on the sea coast. These I believe constitute their principal food, but I have no doubt but that they also feed on flesh."

Fig 15. Captain Clark

In 1855, Taylor found hundreds of condors feeding on sea lion carcasses on the California coast. He wrote: "During the early part of the present month, large quantities of sea lions have been killed on the southern coast for the oil; the carcasses of these animals on the beach may be seen at times surrounded by hundreds of the Condors. A friend of ours informed us that he saw a few days ago, as many as three hundred of these creatures near such feeding ground, within a distance of a league." In the 1860s, condors feeding on seal and whale carcasses in California **(Figure 16)** were reported, although it was never directly observed of them doing so.

Fig 16. Condors devouring seals

Documented dietary shifts have an important meaning for understanding the past distribution of condors. The collected data demonstrate marine mammals were an important component of the diet of condors in coastal California during the Pleistocene, even when large terrestrial mammals were relatively abundant. It is highly unlikely Pleistocene condors living in interior regions had marine-dominated diets, as observed for 40 percent of the animals in a sample from the La Brea tar pits. Thus, results support the **hypothesis** that the restriction of the range of condors to the Pacific coast after the Pleistocene mega faunal extinction was largely controlled by the presence of a "fallback" food source, marine mammals, which at least some of the population was already using. The switch to terrestrial foods in historical condor populations may reflect the reduction of **pinnipeds** and whale populations related to commercial hunting in the late 1700s through the early 1900s. At the same time, however, the expansion of cattle ranching in California and elsewhere in the American west offered condors a new source of abundant large terrestrial carcasses that allowed them to shift eastward, away from coastal refuges. Historical accounts of condor feeding patterns in the 1800s show cattle and deer comprised the major component of their diet. Data indicate a combination of range livestock and wild **ungulates** (deer and elk) remain a component of the diet of the birds sampled, but a major portion of their diet was provided by humans in the form of stillborn calves of corn fed cattle.

The development of conservation strategies for viable condor populations requires adequate and safe food supplies exist for these birds in the wild. Coastal regions lack abundant carcasses of large land mammals and, throughout the former and present range of the condors in southern and central California, this food supply is likely to become increasingly scarce. Efforts to establish a self-sustaining condor population may be enhanced, however, by the widespread availability of marine mammals as an additional food source. This strategy is particularly attractive, in that pinniped populations are reestablishing along the coast of California.

As nature's clean-up crew, condors and other carrion eaters often eat organisms, such as dead and decaying animals that are harmful to humans and the environment. A condor may eat up to 3 to 4 pounds at a time. They may not need to eat again for several days. The condors' exceptionally **corrosive** stomach acids, allow them to safely digest putrid carcasses infected with botulinum toxin, hog cholera bacteria and anthrax bacteria that would be lethal to other scavengers, thus, removing these bacteria from the environment. Condors, as well as other vultures, sometimes vomit when threatened or approached. Contrary to some accounts, they do not "projectile vomit" on their attacker as a deliberate defense but it does lighten their stomach load to make take-off easier and the vomited meal residue may distract a predator, allowing the bird to escape. **They help keep us safe and the environment clean!** Condors like to be clean, too. In fact, it is important for all birds to keep their feathers neat and well-groomed; but you've never seen a bird with a hair brush, right? Instead, they use their beaks to clean or preen their feathers.

Chapter 4

Reproduction

When a condor reaches six or seven years of age, he or she is ready to find a mate. To attract a prospective mate, the male condor performs a display **(Figure 17)**, in which the male turns his head red and puffs out his neck feathers. He then spreads his wings and slowly approaches the female. If the female lowers her head to accept the male, the condors become mates for life. Condors will breed once every other year, with an elaborate courtship flight and dance leading up to mating. Before that, an appropriate nest site must be found.

Fig 17. Courtship display

The pair makes a simple nest in caves or on cliff clefts, especially ones with nearby roosting trees and open spaces for landing. A mated female lays one bluish-white egg **(Figure 18)** every other year. Eggs are laid as early as January to as late as April. The egg weighs about 10 oz. and measures from 3.5 to 4.7 in. long and about 2.6 in. wide **(Side Bar G)**. If the chick or egg is lost or removed, the parents "double clutch", or lay another egg to take the place of the lost one. Researchers and breeders take advantage of this behavior to double the reproductive rate by taking the first egg away for puppet-rearing; this induces the parents to lay a second egg, which the condors are sometimes allowed to raise.

Side Bar G: Cool Facts about the California condor

- In the late Pleistocene, about 40,000 years ago, California Condors were found throughout North America. At this time, giant mammals roamed the continent, offering condors a reliable food supply. When Lewis and Clark explored the Pacific Northwest in 1805 they found condors there. Until the 1930s, they occurred in the mountains of Baja California.

- One reason California Condor recovery has been slow is their extremely slow reproduction rate. Female condors lay only one egg per nesting attempt, and they don't always nest every year. The young depend on their parents for more than 12 months, and take 6-8 years to reach maturity.

- Condors soar slowly and stably. They average about 30 mph in flight and can get up to over 40 mph. They take about 16 seconds to complete a circle in soaring flight. By comparison, Bald Eagles and Golden Eagles normally circle in 12–14 seconds, and Red-tailed Hawks circle in about 8–10 seconds.

- At carcasses, California Condors dominate other scavengers. The exception is when a Golden Eagle is present. Although the condor weighs about twice as much as an eagle, the superior talons of the eagle command respect.

- Condors can survive 1–2 weeks without eating. When they find a carcass they eat their fill, storing up to 3 pounds of meat in their crop (a part of the esophagus) before they leave.

- California Condors once foraged on offshore islands, visiting mammal and seabird colonies to eat carrion, eggs and possibly live prey such as nestlings.

- In cold weather, condors raise their neck feathers to keep warm. In hot weather, condors (and other vultures) urinate onto a leg. As the waste evaporates, it cools off blood circulating in the leg, lowering the whole body temperature. Condors bathe frequently and this helps avoid buildup of wastes on the legs.

- Adult condors sometimes temporarily restrain an overenthusiastic nestling by placing a foot on its neck and clamping it to the floor. This forceful approach is also a common way for an adult to remove a nestling's bill from its throat at the end of a feeding.

- Young may take months to perfect flight and landings. "Crash" landings have been observed in young four months after their first flight.

- California Condors can probably live to be 60 or more years old—although none of the condors now alive are older than 40 yet.

- What's in a name? The name "condor" comes from *cuntur*, which originated from the Inca name for the Andean Condor. Their scientific name, *Gymnogyps californianus*, comes from the Greek words *gymnos*, meaning naked, and refers to the head, and *gyps* meaning vulture; *californianus* is Latin and refers to the birds' range.

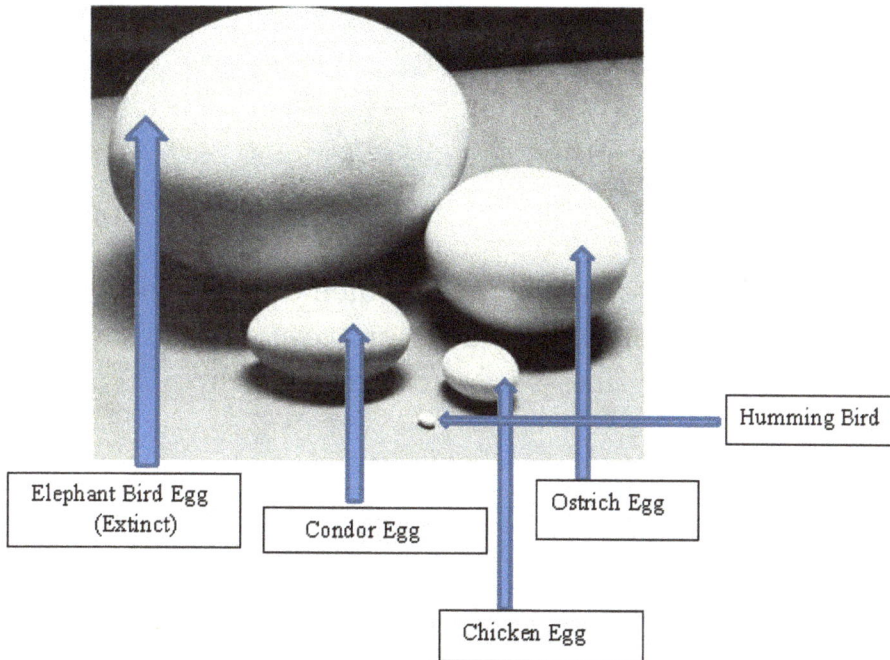

Elephant Bird Egg (Extinct)

Condor Egg

Chicken Egg

Ostrich Egg

Humming Bird

Fig 18. Comparative egg sizes

Raising a condor chick requires tremendous energy and time **(Figure 19)**. The egg is incubated by both parents and hatches after approximately 56 days. Both parents share responsibilities for providing the nestling with food and warmth.

Fig 19. Condor chick

Chicks hatch with their eyes open and sometimes can take up to a week to leave the shell completely. The young are covered with a grayish down **(Figure 20)** until they are almost as large as their parents.

Fig 20. Condor habitat

At two to three months of age, condor chicks leave the nest but remain in the vicinity of the nest where they are fed by their parents. The chick takes its first flight at about six to seven months of age but may not become fully independent until the following year. Chicks learn from their parents how to fly and find food. Parents will look after their young up to two years after hatching.

Though they are able to fly after five to six months, chicks continue to roost and forage with their parents until they are in their second year, at which point the parents typically turn their energies to a new nest. Ravens are the main predatory threat to condor eggs, while golden eagles and bears are potential predators of condor offspring.

Chapter 5

Habitat

The condors live in rocky shrub land, coniferous forests, and oak savannas. They are often found near cliffs **(Figure 21)** or large trees, which they use as nesting sites. Individual birds have a huge range and have been known to travel up to 160 mi. in search of carrion.

Fig 21. Condor with chick

There are two sanctuaries dedicated to this bird, the Sisquoc Condor Sanctuary in the San Rafael Wilderness **(Figure 22)** and the Sespe Condor Sanctuary **(Figure 23)** in the Los Padres National Forest. These areas were chosen because of their prime condor nesting habitat.

Fig 22. San Rafael Wilderness

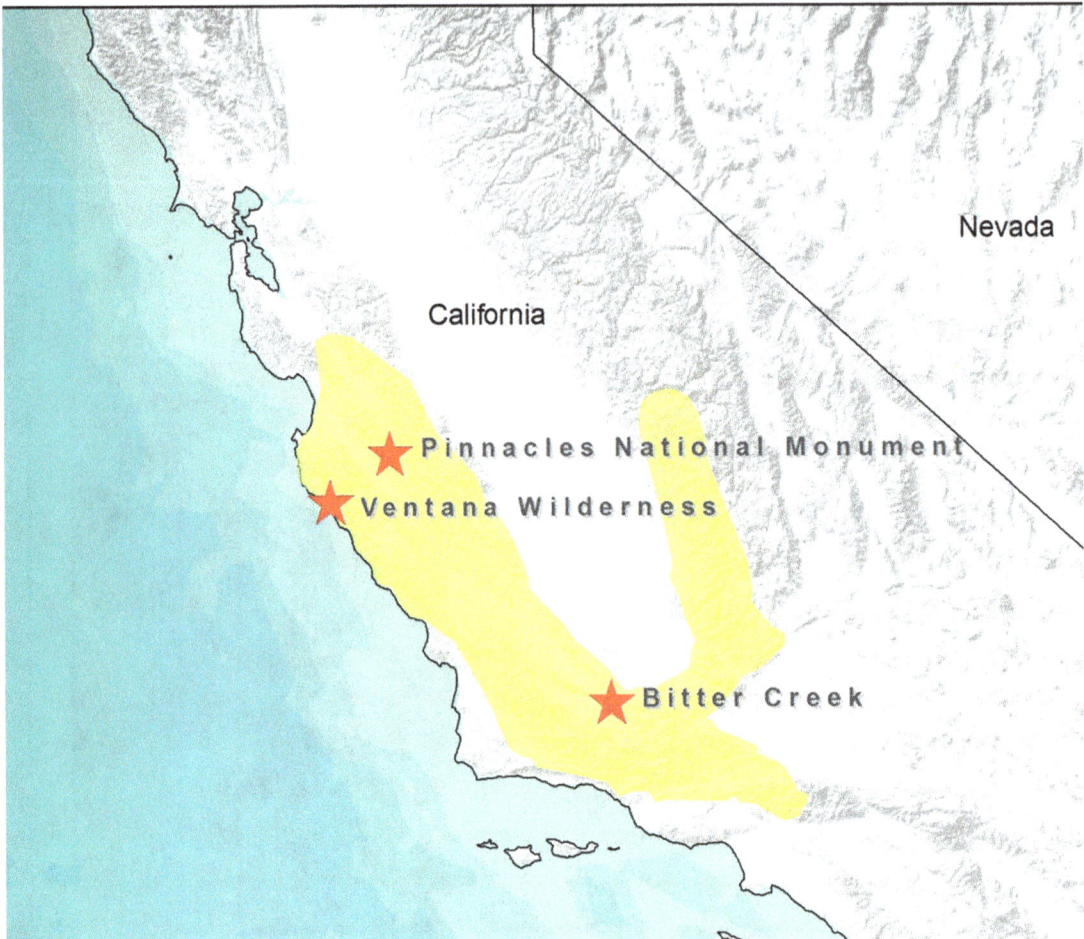

Fig 23. Condors occupy a wishbone-shaped portion of California

Condors occupy a wishbone-shaped portion of California **(Figure 24)** extending from Santa Clara County (rarely San Mateo County) south to Ventura County, then north to Fresno County. This area corresponds roughly with the mountainous terrain surrounding the San Joaquin Valley; the Coast Ranges on the west, Transverse and Tehachapi Mountains at the south, and the Sierra Nevada on the east. Because the condor is a soaring bird, it depends, to some extent, on thermal updrafts and wind currents of mountain terrain for transport. Historically, however, condors traveled far out over the almost flat Central Valley, as much as 40 miles from foothill areas.

Fig 24. Condor nest

Recent restriction by condors to the more mountainous areas of the State is based, to a great extent, on habitat availability. The San Joaquin Valley, formerly grassland with herds of native big game and domestic livestock, now is predominantly intensively managed cropland. The Los Angeles metropolitan area, with 10 million inhabitants, extends along the south boundary of the condor range for 90 miles east and west, and 45 miles north and south. Similarly, the San Francisco Bay area, with 4 million inhabitants and 2,900 square miles of urbanization and industrialization, forms a barrier to the north. Although downtown Los Angeles is less than 45 miles from the principal condor nesting areas in Ventura County, the condor range itself is sparsely populated and, with only a few exceptions, receives limited human use. Thirty-six percent is in public ownership, principally administered by the U.S. Forest Service. Much of the 6.9 million acres of privately owned lands is in large holdings. The largest single holding, The Tejon Ranch in the Tehachapi Mountains, includes 290,000 acres, much of it open livestock range.

The climate of the condor range is semi-arid, with 10-30 inches of precipitation, falling mainly as rain between November and May. Snowfalls are usually light and of short duration. No rain falls over much of the area in summer and early fall. Annual temperatures vary from 32 to 109o F. One-half or more of the days during the year are essentially cloudless.

Most condor habitat is located between 1,000-9,000 feet elevation. This includes all of the Coast Range and Tehachapi Mountains, but only the western slopes of the Sierra Nevada. Distance from food sources may preclude regular use of higher elevations. Condor nesting occurs in the area 2,000-4,500 feet above sea level.

The nesting areas are characterized by extremely steep, rugged terrain, with dense brush surrounding high sandstone cliffs **(Figure 25)**. Principal plant species are several types of Ceanothus, live oaks, chamise, silk tassel bush, and poison oak. Interspersed with the brush are small groves of big cone Douglas-fir, which are favored roosting areas for condors. The cliffs have numerous crevices and wind- and water-created caves, in which condors lay their eggs.

Fig 25. Sespe Condor Sanctuary

Within the brush land (chaparral) community, there are small openings (potreros) dominated by annual grasses. In the Coast Ranges, through the Tehachapi Mountains, and in the foothills of the Sierra Nevada are vast areas of open grassland dominated by introduced annual grasses, particularly wild oats and cheat grass. Some stretches are almost treeless; others have scatterings of oaks, walnuts, and related trees. In these open areas occupied by domestic livestock, condors find most of their food supply. Although all nesting sites are within the Los Padres National Forest, almost all condor feeding areas are privately owned. In the higher portions of the Transverse Ranges, and above about 6,000 feet in the Sierra Nevada, are stands of several species of conifers. These forest areas are occupied by nonbreeding condors as summer roosting sites and the open rangelands below provide food for them.

Chapter 6

Threats

Causes of condor decline have been diverse and difficult to document. It appears most are related to mortality factors, such as poisoning, shooting and collisions with power lines, rather than reproductive failure. Records suggest nesting success of the condor over the last 40 years has been about 50 percent, which compares favorably with several other species of vultures not endangered. The use of pesticides and other poisons in California certainly has contributed to condor mortality. Because it feeds on carcasses, the condor often ingests the poisons that killed the prey, such as DDT, cyanide, or strychnine. Condors have been known to suffer from lead poisoning after ingesting pellets from animals killed by hunters **(Figure 26)**. Levels of ingested poisons may not be fatal to adults but will kill chicks and immature birds.

Fig 26. Oil pumps

Studies in San Diego Zoo Disease Investigations show the main obstacle preventing condor populations from becoming **viable** in the wild is lead poisoning, resulting from ingestion of ammunition fragments in carcasses left by hunters. Ingested lead is absorbed into the blood stream, causing gastrointestinal and nervous system problems that can result in debilitation and death if left untreated. In an attempt to reduce the exposure of condors to lead, the field team provides lead-free food subsidies to the condors at all release sites. Routine feeding has the added benefit of giving researchers opportunities for periodic re-trapping of wild individuals to conduct health inspections and treat condors for lead exposure, when necessary.

Threats to California condors are many. According to IUCN (**International Union for Conservation of Nature),** they include:

- **Energy production and mining:**

  Renewable energy - includes threats associated with exploring, developing and producing non-living resources. Within this category are threats associated with:

  a. Oil and gas drilling **(Figure 27),**

  b. Mining and quarrying **(Figure 28),**

  c. Renewable energy (geothermal, solar, wind, tidal; **Figure 29).**

- **Transportation and service corridors**

  Utility and service lines

  a. Roads and railroads **(Figure 30),**

  b. Utility and service lines **(Figure 31),**

  c. Shipping lanes **(Figure 32),**

  d. Flight paths **(Figure 33),**

- **Biological resource use**

  Hunting and trapping terrestrial animals **(Figure 34)**

The 'biological resource use' category (IUCN 5) includes any "threat of consumptive use of wild biological resources, including the effects of deliberate and unintentional harvesting; including the persecution or control of specific species". The types of biological resource use include: Hunting and collecting of terrestrial animals: This is defined as the killing or trapping of terrestrial wild animals or animal products for commercial, recreation, subsistence, research or cultural purposes, or for control/persecution. This also includes accidental mortality and bycatch.

Fishing and harvesting aquatic resources: This is the harvesting of aquatic wild animals or plants for commercial, recreational, subsistence, research, or cultural purposes, or for control/persecution reasons. This also includes accidental mortality and bycatch.

- **Invasive and other problematic species, genes and diseases**

Viral/prion-induced diseases

Threats from non-native and native plants, animals, pathogens/microbes, or genetic materials that have or are predicted to have harmful effects on biodiversity following their introduction spread and/or increase in abundance:

1. Invasive non-native/alien species/diseases. Harmful plants, animals, pathogens and other microbes not originally found within the ecosystem(s) in question and directly or indirectly introduced and spread into it by human activities

2. Problematic native species/diseases. Harmful plants, animals, or pathogens and other microbes that are originally found within the ecosystem(s) in question but have become out-of-balance or released directly or indirectly because of human activities

3. Introduced genetic material. Human altered or transported organisms or genes

4. Problematic species/diseases of unknown origin

5. Viral/prion-induced diseases

6. Diseases of unknown cause

- **Pollution**

Agricultural and forestry effluents **(Figure 35)**

Pollution threats from introduction of exotic and/or excess materials or energy from point and nonpoint sources:

1. Domestic and urban waste water—Water-borne sewage and non-point runoff from housing and urban areas that include nutrients, toxic chemicals and/or sediments

2. Industrial and military effluents—Water-borne pollutants from industrial and military sources including mining, energy production, and other resource extraction industries that include nutrients, toxic chemicals and/or sediments. Includes oil spills.

3. Agricultural and forestry effluents—Water-borne pollutants from **agriculture**, **silviculture**, and **aquaculture** systems that include nutrients, toxic chemicals and/or sediments including the effects of these pollutants on the site where they are applied. 4. Garbage and solid waste—Rubbish and other solid materials including those that entangle wildlife.

5. Air-borne pollutants—Atmospheric pollutants from point and nonpoint sources.

6. Excess energy—Inputs of heat, sound, or light that disturb wildlife or ecosystems.

In summary, there are many threats to the condor; lead poisoning is the leading cause of death for these slow-to-reproduce scavengers. Condors eat dead animals, including those that have been shot with lead ammunition. Sometimes carcasses or entrails are left behind by hunters who clean their kills in the field, or an animal is shot and gets away but later dies. Lead can quickly accumulate in the big birds, causing lead poisoning leading to anemia, blindness, seizures and death.

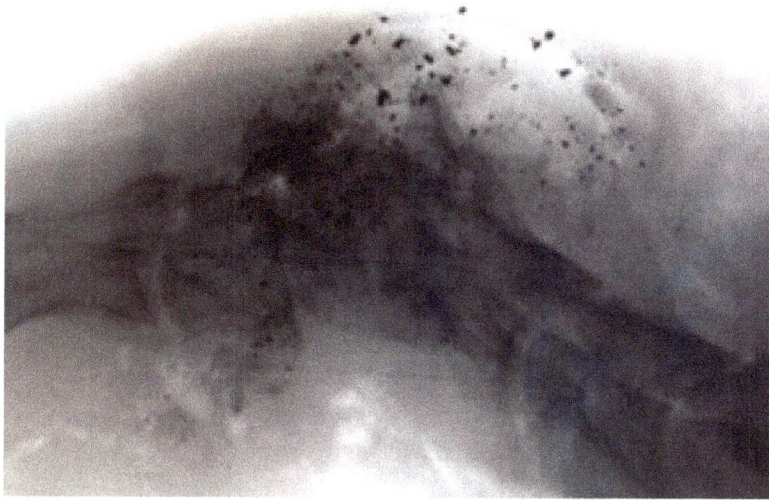

Fig 27. X-ray of deer neck with buckshot

Fig 28. Solar panel field

Fig 29. Mining and quarrying

Fig 30. Railroad tracks-railway

Fig 31. Power Lines

Fig 32. Shipping lanes

Fig 33. Airplane view from Hayden Butte

Fig 34. Cage trap

47

Fig 35. Fish kill

Chapter 7

Recovery

What can be done? **(Side Bar H)** Can we get the condor back? **(Side Bar I)**. The US Fish and Wildlife Service has a plan.

Side Bar H: Facts and What You Can Do to Help

In 1987, there were only 24 California Condors left. All were captured and brought to the San Diego Wild Animal Park and the Los Angeles Zoo for breeding. Now there are nearly 400 living, though they aren't able to sustain their population in the wild and still need breeding facilities.

All California Condors have wing bands so scientists can track them. That is how they know that birds living in different areas actually "visit" each other in the wild. Each condor has a documented history. You can see some of it on the Santa Barbara Zoo Condor website (https://www.sbzoo.org/animal/condor/) .

Lead poisoning is a lead killer of California Condors. Hunters still use lead bullets to hunt animals. Condors eat dead animals, ingest the bullets and die from lead poisoning. California has made it illegal to hunt with lead bullets in condor territory, but some still do anyway. The other reasons for the decline in numbers is loss of habitat, poaching, wind turbines, and power lines.

California Condors can live for up to 50 years.

California Condors are the largest North American land bird with a wingspan of almost 3 meters.

California Condors bred in captivity are taught to avoid power lines and people.

The number one thing that kids (and adults) can do is clean up trash! California Condors gravitate toward micro trash, which are small bits of trash like scraps and bottle caps. Scientists aren't sure why condors like it, but they bring the trash back with food for their chicks. Small bits of trash can block a tiny chick's airway and cause it to die. So if you're in California condor country (they don't fly out of their way to seek it out), dispose of every single shred of trash by packing it away. Don't leave trash out. But hopefully lessons like these will encourage kids not to litter wherever they are.

Did you learn something? You can visit condors in Southern California at the San Diego Zoo, San Diego Wild Animal Park, Lost Angeles Zoo and the Santa Barbara Zoo.

Side Bar I: Condor Recovery Timeline

1937 – 1,200 acres was preserved for the Sisquoc Condor Sanctuary

1947-1951 – 53,000 acres was preserved for Sespe Condor Sanctuary

1967 – California Condors are placed on Federal Endangered Species list

1971 – California Condors are placed on California Endangered Species List

1974 – Hopper Mountain National Wildlife Refuge established 2,471 acres

1982 – *22 condors alive*

1985 – Bitter Creek National Wildlife Refuge established 14,094 acres

1987 – Last wild condor captured 19 Apr using a pit trap at Bitter Creek National Wildlife Refuge

1992 – 8 condors released from Hopper Mountain National Wildlife Refuge

1993 – 5 condors released from Lions Canyon, Sisquoc Condor Sanctuary

1995 – 14 more condors released from Lions Canyon

1996 – 4 condors released from Castle Crags, Machesna Mtn. Wilderness and 4 from Hopper Mtn. NWR

1997 – 4 condors released from Lions Canyon

1999 – 6 condors released from Lions Canyon

2000-2007 – 26 condors released from Hopper Mountain National Wildlife Refuge

2007-2011 – 25 California Condors released from Bitter Creek National Wildlife Refuge

2012 – 33 fledged California Condors are flying wild for the entire program with 16 over Southern California.

June of 2014 – There are 128 condors flying over southern California and a world total of 228 birds flying free, 193 in captivity for a total of …..

421 California condors alive in the world

The primary objective of the California Condor Recovery Plan is reclassification of the California condor to 'threatened' status. The plan provides the criteria for reclassification and outlines the required actions for the accomplishment of each criterion.

The minimum criterion for reclassification to 'threatened' is the maintenance of at least two non-captive populations and one captive population. These populations must:

1. Each number at least 150 individuals,

2. Each contain at least 15 breeding pairs and

3. Be reproductively self-sustaining and have a positive rate of population growth. In addition, the non-captive populations

4. Be spatially separate and non-interacting,

5. Contain individuals descended from each of the 14 founders.

When these five conditions are met, the species should be reclassified to 'threatened' status. The accomplishment of these objectives will depend on reducing mortality to the lowest level possible and ensuring the interchange of individuals among the spatially isolated, free-living sub-populations and the captive flock. We recognized the reestablished condor populations in some areas may require continued artificial feeding to supplement natural food resources and/or to protect birds from exposure to contaminated carcasses; however, such management considerations should not preclude reclassification of the species if the above-listed criteria are met.

Why is reclassification of the condor population to 'threatened' important?

Classification of a species is often confusing. Here are the definitions according to the Endangered Species Act: 'Endangered' means a species is in danger of extinction throughout all or a major portion of its range. 'Threatened' means a species is likely to become endangered within the foreseeable future.

This plan is to move the listing of the condor one step away from endangered. Because the reproduction of condors is a slow process, one egg every other year and the population is so low, under ideal conditions it is not possible to have a population expand so quickly it will become completely out of danger. Therefore, the plan is to increase the condor population one step at a time to complete recovery.

Overall, the recovery was a team effort led in the field by Noel Snyder who was largely responsible for focusing the research. Despite mistakes and setbacks in the early 1980s, leaders persisted and reinvigorated the program by:

1. Adopting a focus on the immediate crisis (reproduction and population growth) of the California condor,

2. Understanding the species and its needs and,

3. Establishing high-performance teams able to experiment and develop effective conservation methods.

Leaders instigated specific new practices, including the first full censuses of wild birds, development of trapping techniques, blood sampling, radio-telemetry, dietary analysis, nest management, captive breeding and release. In 1987, with the remaining wild condors still threatened by lead poisoning, all were taken into captivity for captive breeding, with the aim of releasing the **progeny**. Captive management initially came under the San Diego and Los Angeles Zoos. Dogged determination to continue to develop this process has seen partial and continuing recovery of the species. The population today now exceeds the level of 100 years ago **(Side Bar J)**.

California Condor Population Trends 2010 - 2016

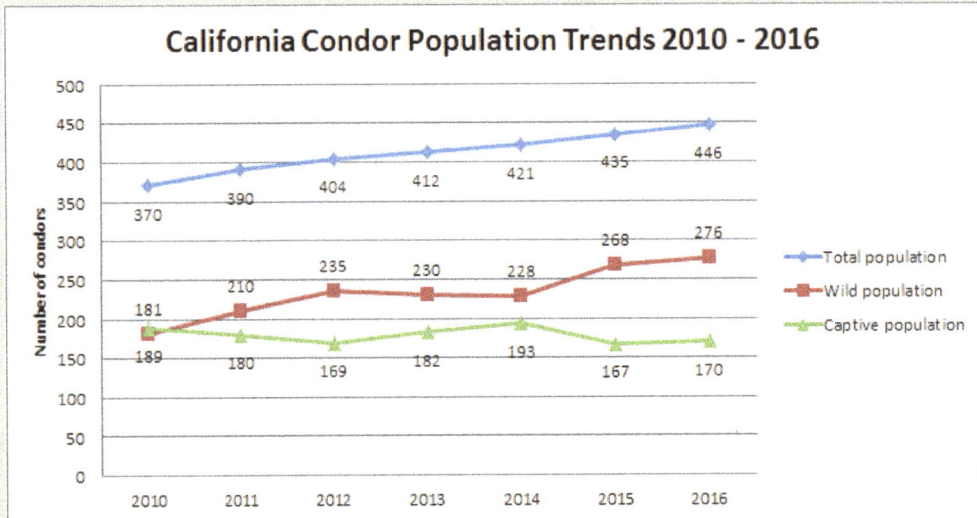

The California Condor Recovery Program (Recovery Program) is a multi-entity effort, led by the U.S. Fish and Wildlife Service, to recover the endangered California condor. Partners in condor recovery include the U.S. Forest Service, National Park Service, Bureau of Land Management, Arizona Game and Fish Department, California Department of Fish and Wildlife, Utah Department of Fish and Wildlife, the federal government of Mexico, the Yurok Tribe, San Diego Zoo, Los Angeles Zoo, Oregon Zoo, Santa Barbara Zoo, Chapultepec Zoo, The Peregrine Fund, Ventana Wildlife Society, and a host of other governmental and non-governmental organizations.

The Recovery Program is now in the final phase of recovery, focusing on the creation of self-sustaining populations. We are placing increased emphasis on the captive-breeding and reintroduction of California condors to the wild and the management of that wild population. These efforts combine trying to reduce the threat of lead with actively managing nesting in the wild to increase the number of wild fledged chicks.

The goal of the California Condor Recovery Plan is to establish two geographically distinct self-sustaining populations, each with 150 birds in the wild and at least 15 breeding pairs, with a third population of condors retained in captivity. As the Recovery Program works toward this goal, the number of release sites has grown. There are three active release sites in California, one in Arizona and one in Baja, Mexico.

As California condors continue to recover, they still face many threats—most from humans. They include:

- **Pollution**

  Human activity poses the greatest threat to condors. These big birds pick up and feed their young small objects left behind by humans, such as pennies and pieces of glass, known as micro trash, which can kill chicks. Condors have been known to drink poisons accidentally, such as antifreeze.

- **Illegal Hunting**

  Some people even shoot condors, though it's been illegal to do so for nearly 100 years.

- **Habitat Loss**

  Human development has degraded or completely destroyed the habitat condors require for foraging, nesting and roosting. Housing developments, oil and gas production and large-scale solar and wind projects can destroy or fragment important condor habitat. With their massive wingspan of nearly 10 feet, power lines pose a major threat of electrocution to condors.

- **Lead poisoning**

  Hunters and poachers sometimes leave their kills behind and that can also mean the lead ammunition often used to kill them. Given that condors forage exclusively on dead animals, they are especially susceptible to lead exposure from carcasses left in the field. Lead poisoning from spent ammunition is the number one cause of death among endangered condors in the wild.

Chapter 8

Future of the Condor

Despite their comeback, California condors remain an endangered species and threatened by human habits. The historical decline of the condors was a result of a high mortality rate paired with a low reproductive rate. This combination still threatens them today. Poisoning, pollution, development of habitat and shooting continue to pose a perilous threat to the species. Condors are also endangered by poaching and collisions with powerlines. Out of these, lead poisoning and micro trash ingestion **(Figure 36)** pose two of the largest threats to condors.

Fig 36. Micro Trash

While California condor populations are rebuilding, these great birds have yet to return to anywhere near their former range, including Oregon. The Oregon Zoo is helping reestablish populations in California and Arizona through their captive breeding effort but the organization states efforts to return condors to Oregon and the Pacific Northwest remains a distant goal; a goal that "will require us to think in terms of the restoration of whole ecosystems, including healthy salmon runs, viable mammal populations, and even the vitality of our own cultures and ability to coexist with and regenerate natural systems". Restoring our ecosystems also means bringing back top predators, such as the gray wolf, that play a vital role in shaping and keeping ecosystems in balance, as well as provide carrion for condors. Oregon Wild is working hard to do just this: fighting to protect the rivers, forests, wilderness, and wildlife that make Oregon such a great place to be.

Hopefully, with time, the work of Oregon Wild, other organizations, and individuals will allow the California condor to spread its wings over the Northwest once again.

Glossary

Agriculture – is the science and art of cultivating plants and livestock.

Albedo – reflective power; the fraction of incident radiation (such as light) that is reflected by a surface or body (such as the moon, earth or a cloud).

Aquaculture – also known as aquafarming, is the farming of fish, crustaceans, molluscs, aquatic plants, algae and other organisms.

Bifurcation – the point or area at which something divides into two branches or parts

Biome – is a community of plants and animals that have common characteristics for the environment in which they exist.

Bronchi – are the main passageway into the lungs

Chronospecies – is a species derived from a sequential development pattern. This sequence of alterations eventually produces a population which is physically, morphologically, and/or genetically distinct from the original ancestors.

Corrosive –wearing away gradually, usually by chemical (acidic) action

Epoch – a division of geologic time less than a period and greater than an age.

Hypothesis – used in scientific research as a tentative assumption made to draw out and test its logical or empirical consequences.

IUCN – is the global authority on the status of the natural world and the measures needed to safeguard it.

Paleontology – is the scientific study of life that existed prior to, and sometimes including, the start of the Holocene **Epoch** (roughly 11,700 years before present).

Pinniped – fin or flipper-footed and refers to the marine mammals that have front and rear flippers. This group includes seals, sea lions, and walruses. These animals live in the ocean but are able to come on land for long periods of time.

Pleistocene Epoch – lasting from about 2 million years ago to 12,000 years ago.

Prehension – the act of taking hold, seizing, or grasping.

Progeny – a descendant or offspring.

Scavengers – are animals that feed on animal carcasses they did not kill.

Silviculture – is the practice of controlling the growth, composition, health, and quality of forests to meet diverse needs and values.

Syringeal vocalizations – bird calls produced through the vocal organ of birds, situated at or near the **bifurcation** of the **trachea** into the **bronchi**.

Relict – a surviving species of an otherwise extinct group of organisms.

Trachea – the main trunk of the system of tubes by which air passes to and from the lungs in vertebrates.

Tundra – is a type of **biome** where the tree growth is hindered by low temperatures and short growing seasons.

Viable – capable of growing or developing

Ungulates – are any members of a diverse group of primarily large mammals that includes   odd-toed ungulates such as horses and rhinoceroses, and even-toed ungulates such as cattle, pigs, giraffes, camels, deer, and hippopotamuses.

Weichsel glaciation – was the last glacial period and its associated glaciation in Northern Europe.

Side Bars

Side Bar A Did you know?

Side Bar B Fun Fact

Side Bar C Geologic Timetable

Side Bar D Quaternary Period

Side Bar E The Pleistocene Epoch—The Rise of the Condor

Side Bar F The Condor Is a Chronospecies

Side Bar G Cool Facts about the California condor

Side Bar H Facts and What You Can Do to Help

Side Bar I Condor Recovery Timeline

Side Bar J California condor Population Trends 2010 – 2016

List of Illustrations

Fig. 1 California Condor in Flight

Fig. 2 White beneath wing

Fig. 3 Head and Neck

Fig. 4 Size comparison

Fig. 5 Condor Foot

Fig. 6 Ice Age Glacier

Fig. 7 Europe during the last glacial period

Fig. 8 Ice Sheet

Fig. 9 55 million years of falling temperature

Fig. 10 Climate Change in the Pleistocene

Fig. 11 Ice melt

Fig. 12 Sheep as carrion

Fig. 13 Condor feeding on carrion

Fig. 14 Marine mammal carcass

Fig. 15 Captain Clark

Fig. 16 Condors devouring seal

Fig. 17 Courtship display

Fig. 18 Comparative egg sizes

Fig. 19 Condor chick

Fig. 20 Condor habitat

Fig. 21 Condor with chick

Fig. 22 San Rafael Wilderness

Fig. 23 Condors occupy a wishbone-shaped portion of California

Fig. 24 Condor nest

Fig. 25 Sespe Condor Sanctuary

Fig. 26 Oil pumps

Fig. 27 X-ray of deer neck with buckshot

Fig. 28 Solar panel field

Fig. 29 Mining and Quarrying

Fig. 30 Railroad tracks-railway

Fig. 31 Power Lines

Fig. 32 Container Ship Freighter

Fig. 33 Airplane view from Hayden Butte

Fig. 34 Cage trap

Fig. 35 Fish kill

Fig. 36 Micro trash

Illustration Credits

https://en.wikipedia.org/wiki/California_condor

https://www.wildlife.ca.gov/conservation/birds/california-condor

https://www.oregonzoo.org/discover/animals/california-condor

https://www.peregrinefund.org/projects/california-condor

https://digitalmedia.fws.gov/digital/collection/natdiglib/id/14119/rec/5

Dirk Beyer - Own work, CC BY-SA 3.0, https://commons.wikimedia.org/w/index.php?curid=352940

https://en.wikipedia.org/wiki/Chronospecies

https://www.smithsonianmag.com/science-nature/mammoths-and-mastodons-all-american-monsters-8898672/

https://commons.wikimedia.org/w/index.php?curid=24341671

https://www.nationalgeographic.org/encyclopedia/ice-sheet/

Olga Ernst - Own work, CC BY-SA 4.0, https://commons.wikimedia.org/w/index.php?curid=71916555

http://www.naturalhistorymag.com/perspectives/082655/condors-and-carcasses

http://www.condortales.com/california-condor/bird-and-mammal-activity-at.html

https://www.fs.usda.gov/recarea/lpnf/recarea/?recid=71868

https://lajollamom.com/california-condor-california-state-bird/

https://www.wildlife.ca.gov/conservation/birds/california-**condor**

https://www.sfgate.com/bayarea/article/First-condor-nest-in-Pinnacles-in-100-years-3270832.php

https://www.theatlantic.com/photo/2014/08/the-urban-oil-fields-of-los-angeles/100799/

https://www.sccgov.org/sites/dpd/DocsForms/Documents/Permanente_20141031_MRRC.pdf

https://unsplash.com/search/photos/solar-farm

https://www.pexels.com/search/**railroad**/

https://www.popsci.com/why-dont-we-put-power-lines-underground/

https://pixabay.com/photos/container-container-ship-freighter-3397074/

http://www.bryondarby.com/photos/flight-paths/

https://en.wikipedia.org/wiki/Trapping

Quaggiotto, Maria Martina. 2016. The role of marine mammal carrion in the ecology of coastal systems. PhD thesis. http://theses.gla.ac.uk/7099/

Ulamm - File:Europe topography map.png, 2 April 2006 by San Jose, based on the Generic Mapping Tools and ETOPO2, CC BY-SA 3.0,

Sources

https://en.wikipedia.org/wiki/California_condor

https://defenders.org/california-condor/basic-facts

https://animals.sandiegozoo.org/animals/california-condor

https://institute.sandiegozoo.org/species/california-condor

https://www.wildlife.ca.gov/conservation/birds/california-condor

https://www.nationalgeographic.com/animals/birds/c/california-condor/

https://www.fws.gov/cno/es/calcondor/condor.cfm

https://www.peregrinefund.org/explore-raptors-species/California_Condor

https://www.lazoo.org/blog/2016/09/14/conservation-corner-california-condor-recovery-program/

https://oregonwild.org/wildlife/california-condor

https://www.britannica.com/science/Pleistocene-Epoch

http://www.dandebat.dk/eng-klima5.htm

https://en.wikipedia.org/wiki/Paleontology

https://en.wikipedia.org/wiki/Chronospecies

https://www.allaboutbirds.org/guide/California_Condor/lifehistory

http://www.friendsofcondors.org/during-the-pl

https://www.ncbi.nlm.nih.gov/pmc/articles/PMC1283853/

http://www.condortales.com/california-condor/bird-and-mammal-activity-at.html

https://lajollamom.com/california-condor-california-state-bird/

http://www.naturalhistorymag.com/perspectives/082655/condors-and-carcasses

https://oregonwild.org/wildlife/california-condor

https://www.fws.gov/cno/es/calcondor/CondorCount.cfm

https://www.networx.com/article/landscaping-native-plants-vs-non-nativ

https://www.networx.com/article/habitat-fragmentation-is-for-the-birds

*Church, ME; Gwiazda, R; Risebrough, RW; Sorenson, K; Chamberlain, CP; Farry, S; Heinrich, W; Rideout, BA; Smith, DR. 2006. "Ammunition is the Principal Source of Lead Accumulated by California Condors Re-Introduced to the Wild". Environmental Science & Technology. **40** (19): 6143–50.*

Snyder, Noel and Snyder, Helen. 2000. The California Condor. **University of California Press, Berkley, CA.**

West, Christopher J. 2009. Factors influencing vigilance while feeding in reintroduced California condors (*Gymnogyps californianus*). MS Thesis, Humboldt State University, Acadia, CA.

Wilbur, Sanford R. 1978. The California Condor, 1966-76: A Look at its Past and Future. North American Fauna, Number 72, US Fish and Wildlife Service, Washington, DC.

About the Author

Mary Jo Nickum is a retired science librarian. She has a B. A. in English education, a Master's in librarianship and an interdisciplinary Master's in fisheries science, anthropology and education. Her first book for children was a chapterbook, *Mom's Story, a Child Learns about MS.* She specializes in writing science for children. Ms. Nickum is an award winning author for her books and science articles.

Mary Jo lives in Fountain Hills, Arizona with her husband. She has two grown sons.

www.ingramcontent.com/pod-product-compliance
Lightning Source LLC
Chambersburg PA
CBHW080629030426
42336CB00018B/3128